os and Cons of
COAL, GAS, and OIL

Sally Morgan

rosen publishing's
rosen
central

New York

Published in 2008 by The Rosen Publishing Group, Inc.
29 East 21st Street, New York, NY 10010

First Edition

Series Editor: Jennifer Schofield
Editor: Debbie Foy
Consultant: Rob Bowden
Designer: Jane Hawkins
Cover designer: Paul Cherrill
Picture Researcher: Diana Morris
Illustrator: Ian Thompson
Indexer: Sue Lightfoot

Picture Acknowledgments:
Jean Pierre Amet/Sygma/Corbis: 20. Remi Benali/Corbis: 15.
Jon Bower/Ecoscene: 9. Andrew Brown/Ecoscene: 25.
EASI-Images: front cover br, 7, 36, 39,45. Richard Folwell/SPL:
front cover, 17. Chinch Gryniewicz/Ecoscene: 34. HIP/Topfoto: 4.
Peter Hulme/Ecoscene: 1, 12. Alexandra Elliott Jones/Ecoscene: 27.
Chris Knapton/Ecoscene: 18. Reuters/Corbis: 40. Erik Schaffer/
Ecoscene: 22. Keren Su/Corbis: 30. Peter Turnley/Corbis: 43.
Visual & Written/Ecoscene: 33. Jim Winkley/Ecoscene: 10.
David Wootton/Ecoscene: 28.

Library of Congress Cataloging-in-Publication Data

Morgan, Sally.
  the pros and cons of Coal, Gas, and Oil / Sally Morgan. -- 1st ed.
      p. cm. -- (The energy debate)
  Includes index.
  ISBN-13: 978-1-4042-3744-5 (library bdg.)
  ISBN-10: 1-4042-3744-5 (library bdg.)
 1. Fossil fuels. 2. Fossil fuels--Economic aspects. 3. Petroleum
reserves. I. Title.
  TP318.M67 2007
  333.8'2--dc22

                                    2006038675

Manufactured in China

Contents

CHAPTER 1   Coal, gas, and oil and the energy debate

The twenty-first century world is hungry for energy. Vast quantities of electricity are generated to power industry and homes; billions of gallons of oil are used for transportation.

Twenty-first century energy

The modern Western home is filled with a range of electrical equipment, such as televisions, DVD players, computers, and kitchen appliances. There are lights in all the rooms and a supply of hot water for heating and washing. These all use energy.

The world's population is around 6.5 billion and it is increasing rapidly, especially in less economically developed countries (LEDCs). As the number of people increases, the use of energy increases, too. A person living in an LEDC uses much less energy than somebody living in Europe or North America. For example, someone in the United States uses up to 30

▽ Chimneys pumping out clouds of dark, air-polluting smoke were a common sight in the industrial areas of Victorian Britain.

times more fossil fuel energy than someone who lives in an LEDC, such as Ethiopia or Tanzania. However, as standards of living improve in many LEDCs, more people have the means to buy goods, such as refrigerators, televisions, computers, and cars. The economies of large countries, such as India and China, are growing rapidly, and this is increasing the demand for electricity and oil.

Burning fossil fuels
The main sources of energy in the world are fossil fuels—coal, gas, and oil. However, the use of these fuels causes massive environmental problems. When fossil fuels are burned, they release several gases. One such gas is carbon dioxide (CO_2). CO_2 is one of the gases in the atmosphere that is responsible for global warming.

▽ This pie chart shows the energy sources used in 2005 to generate the world's electricity supply.

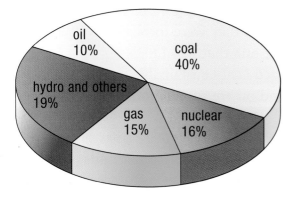

oil 10%
coal 40%
hydro and others 19%
gas 15%
nuclear 16%

❝We have to economize, be more efficient, and move away from fossil fuels. Renewable energy sources, such as wind, wave, and solar, have a role to play, but will not really come into their own until we have a way of storing their energy.❞

Lord Ronald Oxburgh, former chairman, Royal Dutch Shell

Global warming is causing the average temperature of the Earth's atmosphere to rise. Other gases produced by burning fossil fuels include sulphur dioxide and nitrogen oxides, which cause acid rain. Acid rain is rain water with a pH value lower than 5.6. Normally, water has a neutral pH. The acidity of this rain water comes from sulphur dioxide and nitrogen oxides, which dissolve in water in the air to form sulphuric and nitric acids.

Most of the world's leaders recognize that the problem of global warming has to be tackled by the reduction of CO_2 emissions. So, now we are faced with finding ways of reducing our dependency on fuels that emit CO_2, and also ways of supplying our energy needs without creating more CO_2.

Major source of energy

Fossil fuels are given this name because they have formed over millions of years from the decaying remains of plants and animals. Currently, fossil fuels are the world's main source of energy, especially in more economically developed countries (MEDCs). Fossil fuels are used by power stations to generate electricity. They power the engines of cars and airplanes. Substantial quantities of oil are also used in farming and industry. Oil is used to power tractors, harvesters, and other agricultural equipment, and in the manufacturing of agrochemicals, such as nitrogen fertilizers.

Ancient fuels

For several thousand years, people used renewable energy sources, such as the Sun, biomass (wood, leaves, and dung), water, and wind power. Sailing ships and windmills were built to harness wind energy, and energy from water was used to drive water mills. Wood and dung were burned to provide heat and water. Solar energy has been used for thousands of years for drying crops and heating water and buildings. However, by the late sixteenth century, many of the forests in European countries, such as England, had been felled and the supply of wood was severely depleted. An alternative source of energy was needed—and that was coal.

Industrial Revolution

Although coal had been used in parts of the world for hundreds of years, it was not until the eighteenth century that inventors found new ways of using coal as a source of energy. This time of invention started an important period in history known as the Industrial Revolution, which lasted for the next 200 years or so.

During the Industrial Revolution, coal was burned to generate steam, which was used to power machines, such as the steam engine. Steam engines powered the first factories, steamboats, and locomotives. Consequently, the use of coal to generate steam increased greatly. The next step forward was the invention of the internal combustion engine and the car, which used oil instead of coal.

Since then, the use of coal, oil, and gas has steadily increased. However, this has created a problem, because people are using fossil fuels at a rate many times faster than the rate at which they were formed. As a result, the rate that fossil fuels are being used is not sustainable, and coal, gas, and oil are running out.

▷ The vast majority of the global oil supply is used to fuel the steadily increasing number of vehicles that create congestion and pollution on city streets all over the world.

CHAPTER 2 — What Are fossil fuels?

Fossil fuels are formed from the remains of plants and animals that died up to 500 million years ago. Their remains were slowly buried under layers of mud and silt called *sediment*, and then compressed by the weight of this sediment. Over millions of years, the compression turned the decaying plants and animals into coal, natural gas, and oil. Other fossil fuels include peat and *oil shale*, which is a type of rock rich in carbon.

Carbon-rich fuels

Fossil fuels contain the element carbon. Carbon-rich fuels are useful, because they burn in air and release a lot of energy. When they burn, the carbon reacts with oxygen in the air to form CO_2 and water. However, fossil fuels release other gases, such as sulphur dioxide, nitrogen oxides, carbon monoxide and soot, which pollute the environment.

Coal

A hard, brown or black substance, coal provides much of the energy used to generate electricity. It burns with a yellow flame that leaves ash behind. Different types of coal release differing amounts of energy when burned. Coal cannot be used in vehicles, but it can be converted into gases or liquid fuels.

Oil

Oil is a thick liquid that is usually black, but it can be dark green or almost clear. It is made up of many different substances that have to be separated into fractions at an oil refinery (see page 16). The most sought-after fractions are gasoline, diesel, and kerosene. Much of the world's oil production is used in engines of various vehicles, such as cars, trains, boats, and airplanes.

> **"**Our children and grandchildren are going to be mad at us for burning all this oil. It took the Earth 500 million years to create the stuff we're burning in 200 years. Renewable energy sources are where we need to be headed.**"**
>
> Jack Edwards, professor of geology, University of Colorado

Natural gas

Natural gas is made up mostly of methane and is highly flammable. It will cause explosions if sufficient quantities escape and are ignited. Natural gas has no odor, so before it is sent to the pipelines and storage tanks, a chemical is added to give the gas a strong artificial smell. The chemical makes the gas smell like rotten eggs, which means that leaks can be detected more easily.

Natural gas is used mostly for heating and cooking purposes in domestic households and for powering industrial production, especially in manufacturing. Other consumers of gas include glass manufacture, metal refining, and the manufacture of wood and gasolineeum products.

▽ Coal briquettes for sale on the streets of Beijing, China, where coal is a common source of fuel for use in domestic stoves.

Natural gas is a clean-burning fuel, which means that it produces less air pollution than either coal or oil, and can be piped to where it is needed.

The effects of CO_2

A side effect of burning fossil fuels is the release of CO_2. This gas is described as a "greenhouse" gas, since it traps heat in the atmosphere, just like a greenhouse in the yard traps heat from the Sun. As more fossil fuels are burned, more CO_2 is released. The increased levels of CO_2 trap the heat that is released, causing the average temperature of the Earth's troposphere (the lowest part of the atmosphere) to rise. Global warming is now considered a major cause of climate change.

Coal formation

Most of the Earth's coal deposits were formed about 300 million years ago. This period in ancient history was known as the *Carboniferous Period*, because carbon is the basic element found in coal and the other fossil fuels formed at this time.

During this period, the planet's climate was very different—it was much warmer and more humid. There were huge swamp forests with coniferous trees and tree ferns. The seas and lakes were filled with green algae.

Usually, when trees and other plants die, their remains fall to the ground and are broken down by simple microorganisms called decomposers.

△ Most coal-fired power stations are linked to railroad networks, so that quantities of coal can be transported by rail from mines.

However, this did not happen in the swamp forests. When the trees died, they fell into the mud and were buried. There was no oxygen in the mud so the microorganisms could not break down the tree remains. Instead, thick layers of dead plant matter built up in the mud.

Constant pressure

In time, a thick layer of mud or sediment buried the plant remains. The weight of the sediment pushed down on the remains, compressing or squashing them. Water was squeezed out of the remains, and layers of a spongy material called peat were formed. Over millions of years, the continual pressure squeezed out even more water, and the original layer of peat was squeezed to just one-fifth of its original thickness. The peat had been changed into brown coal (lignite).

Over many more years, the brown coal slowly converted into a type of black coal called bituminous coal. More pressure and heat from the Earth turned bituminous coal into anthracite. *Anthracite* is a top-quality coal that is dense, hard, black, and shiny. It has a high concentration of carbon, which gives it a higher energy content than bituminous coal.

Coal is found in layers, which are called seams. A single coal seam can be many feet (meters) thick. Some seams lie close to the surface or are even visible from above ground, and others lie deep underground. Sometimes there are several seams, one above another.

Where is coal found?

Coal is found in around 50 countries in the world. More than 4.3 billion tons (4.4 billion tonnes) of coal were produced in 2004, a 38 percent increase over the previous 20 years. Production of coal has grown fastest in Asia, while it has declined in Europe. The major coal producers are China, United States, Australia, India, and South Africa.

CASE STUDY: Coal on the Rhine

Germany is the largest coal producer in the European Union (EU) and the seventh largest in the world. Germany's recoverable reserves are about 0.7 percent of global reserves. The main coal fields are located close to the River Rhine and around Leipzig in the east, but much of the coal is lignite, a poor-quality brown coal. Germany's rich coal fields, together with large deposits of iron ore, fueled industrial growth in the late nineteenth century, especially in the Ruhr. The Rhine provided cheap transportation, encouraging exports and industrial growth. Today, 75 percent of coal production is used in generating electricity. Coal production has fallen since the early 1990s from over 325 million tons (330 million tonnes), to about 246 million tons (250 million tonnes) in 2003. This has been due to changes to the coal industry after the unification of West and East Germany, and also due to a switch to natural gas.

Coal mining

Hundreds of years ago, coal was dug from cliff faces or shallow trenches. As the demand for coal grew, people dug tunnels into hillsides following the coal seam. These tunnels were called *drift mines*. During the Industrial Revolution, the demand for coal was greater and more efficient methods of extracting coal were needed. In 1700, about 2.95 million tons (3 million tonnes) of coal were mined each year. By 1900, this had increased to 271 million tons (275 million tonnes). A hundred years later, world coal production had risen to over 4.3 billion tons (4.4 billion tonnes). This represented about 24 percent of the world's energy consumption.

> **66** Coal will be providing a substantial proportion of the U.K.'s fuel mix for the generation of electricity in the next decade. Generator demand demonstrates that there has never been a better time to reopen the Hatfield colliery. **99**
>
> Richard Budge, Chief Executive, Powerfuel U.K.

▽ An open-pit coal mine in Scotland. Diggers excavate the coal from the ground, then large trucks carry it to the power station's crushing plant, where it is crushed to a powder.

Open-pit mining

The type of coal mine depends on the depth and quality of the seams and the local geology, as well as environmental factors of the area, such as the need to protect wildlife habitats or avoiding the pollution of water supplies.

In some areas, the coal seams occur close to the surface. This makes it possible to dig out the coal in an open-pit mine. The earth above the coal seam (overburden) is removed, and the coal is dug out using large machines. When the coal is extracted, the overburden is used to fill the hole.

Deep coal mines

Most coal seams are too deep for open-pit mining. Instead, a shaft is dug down to the coal seam, which can be hundreds of feet (meters) below the surface. When the shaft reaches the coal seam, tunnels are dug along the seam and the coal is removed. This type of mining can be dangerous, and miners have to face many dangers, such as tunnel collapses, flooding, and the buildup of poisonous gases. Deep mines produce a lot of waste material, such as rock and soil, which is dumped above ground, creating unsightly spoil heaps.

Today, the two main types of deep mine are room-and-pillar and longwall. In room-and-pillar mining,

THE ARGUMENT: Coal is a valuable source of energy

For:
- There is a plentiful supply in many parts of the world.
- It releases a lot of energy per ton (tonne).
- Coal mining creates jobs.
- Coal is cheaper than oil and gas.

Against:
- Coal, especially lignite, releases polluting gases when it is burned.
- It is bulky to transport, compared to oil and gas.
- Open-pit mining creates environmental damage.

the coal is mined by cutting a series of "rooms" into the coal seam, and leaving behind columns or "pillars" of coal to support the roof of the mine. However, up to 40 percent of the coal is left behind during this method. In longwall mining, a large machine cuts the coal and holds up the roof. The coal falls on a conveyor belt and is removed. When the machine is moved forward, the roof is left to collapse. This method removes 75 percent of the coal.

Oil and gas formation

Oil has been used as a fuel for more than 5,000 years. The Ancient Sumerians, Assyrians, and Babylonians used crude oil and asphalt that leaked out of the ground. The Ancient Egyptians used oil as a medicine for wounds. In North America, Native Americans used blankets to skim oil off the surface of streams and lakes. They used the oil in medicines and to make their canoes waterproof.

From the sea

Oil and gas is formed from the remains of microscopic animals and plants that lived in the sea more than 300 million years ago. When these organisms died, their remains sank to the sea floor, and became covered in a layer of mud and silt. Over millions of years, these layers of sediment built up. As more sediment dropped to the sea floor the remains were compressed and formed oil and natural gas. Bubbles of oil and gas seeped through the rock, filling the tiny spaces around the particles of rock. Eventually, the oil and gas reached a layer of impervious (nonporous) rock through which they could not pass, and pockets of oil and gas were formed.

Crude oil

Crude oil is the term used for the "unprocessed" oil that comes out of the ground. Both crude oil and natural gas are made up of many different substances. For example, crude oil contains substances such as diesel, kerosene, and lubricating oils, but natural gas is mostly methane with small amounts of gases, such as ethane, propane, butane, and pentane. The composition of both crude oil and natural gas can vary widely, however.

THE ARGUMENT: There are many benefits to using oil as fuel

For:

- Oil is a liquid, so it can be pumped from the ground and transported to the refinery.
- Oil contains many different substances, so it has many uses.
- It burns more cleanly than coal.

Against:

- Oil spills harm the environment and wildlife.
- When it is burned, oil releases CO_2 and nitrogen oxides.
- Oil produces more pollution when it is burned, compared to natural gas.
- Oil is flammable, so it can be dangerous to transport and use.

Where is oil and gas found?

The world's largest oil-producing countries are Saudi Arabia, Russia, the United States, Iran, and Mexico. The Middle East, South America, Southeast Asia, the North Sea (between the United Kingdom and Norway), and West Africa are other oil-rich regions.

Natural gas is usually found close to oil. The world's largest producer of gas is Russia. Other major producers include Canada, the United States, the United Kingdom, and Mexico. Gas is also found near coal beds, providing a relatively cheap energy supply since the beds are usually close to the surface, which makes drilling easier. Nowadays, this gas is collected from coal mines and used as a fuel. Since the 1960s, there has been a dramatic increase in the discovery of gas fields, making gas the fastest-growing energy resource. The present global use of natural gas is approximately 20 percent of all fossil fuel use, and this figure is predicted to rise in the future.

▽ This miner is showing what crude oil looks like when it comes straight out of the ground. Crude oil is often mixed with sand, water, and gases.

Extracting and refining oil

For many thousands of years, people collected oil from the surface of the Earth. That changed in 1859, when Edwin Drake dug a well and struck oil in Titusville, Pennsylvania. He figured out how to pump oil to the surface and stored it in wooden barrels. A similar method is used today.

Rotary drilling

Today, the most widely used method of drilling for oil is called *rotary drilling*. The drill and other equipment is supported by a derrick (or drilling rig). The rotary drill works a bit like a screwdriver, turning rapidly to force its way through the ground. The end of the drill is called the bit. This is attached to a drill collar, which in turn is attached to the drill pipe. As the bit turns deeper into the ground, new sections of drill pipe are added.

Drilling fluid is pumped down the inside of the pipe to help break up the rocks and lubricate the drill bit. The drilling fluid and debris (waste) are brought back to the surface outside the drill pipe. Once the drill bit reaches the reserves of oil, the drill pipe is raised and the oil is extracted.

Until the 1970s, most oil wells were vertical. However, new technologies allowed the construction of angled wells, and even horizontal ones.

Horizontal wells are very useful, because they can pass along a reservoir of oil, rather than through it. This means that more oil can be extracted. Angled wells allow engineers to drill for oil that lies below environmentally sensitive areas, such as wildlife habitats or areas of natural beauty. This is called extended reach drilling.

Refining oil

The crude oil that is pumped straight from the ground contains many different compounds, such as gasoline, diesel, kerosene, gas oil, lubricating oils, and fuel oils, and these compounds vary in thickness.

After the crude oil is transported by pipeline or tanker to an oil refinery, the different compounds are separated.

Dissolved gases make up the thinnest oils, and asphalt is the thickest. The most common way of separating the different fractions of oil is through the process of distillation. During this process, the oil is heated. At certain temperatures, the different compounds boil and become vapors, which are then collected and condensed. Kerosene boils at 350°F–500°F (180°C– 260°C), but fuel oil typically boils at higher temperatures—in excess of 626°F (330°C). Some of the fractions of oil may be chemically treated to make other compounds.

CASE STUDY:
North Sea oil

Crude oil was discovered under the North Sea during the 1960s. However, the oil fields were located in the most remote Northern areas, where there was deep water and very rough seas in winter. Due to the high costs of the technology involved, it did not become economic to extract oil from the North Sea until oil prices rose during the late 1970s.

Since then, the North Sea oil fields have made the U.K. and Norway two of the world's major oil-producing countries, boosting their economies and reducing their dependence on oil from the Middle East.

△ An oil production platform, North Sea, which is operated by the Occidental Oil company.

The difficult sea conditions meant that new engineering techniques had to be developed. Costs are high; for example, in 2006 it cost up to $46 million to build and install a new oil platform in the North Sea. Onshore wells cost from $900,000 to $14 million. There are more than 100 oil fields in operation in the North Sea; many are small oil fields that have been discovered in the last ten years. Most of the oil reserves belong to the U.K. and Norway, but some belong to Denmark, Germany, and the Netherlands.

CHAPTER 3 | Harnessing energy from fossil fuels

It is the carbon in fossil fuels that makes them such valuable sources of energy. Most of the world's coal production is used to generate electricity. However, natural gas is now replacing coal as the preferred fuel, since it burns cleaner than coal, which means that it produces less air pollution. For example, natural gas produces between 40 and 50 percent less CO_2 for the same amount of electricity generated as coal.

▽ Steam billows from the cooling towers of this coal-fired power station, which is fitted with a flue gas desulphurization unit.

Inside a coal-fired power station

The coal is transported to the power station, where it is pulveri ed into a powder and blown into a boiler. The burning coal heats water that circulates in pipes around the boiler. The water is then turned into steam, which spins the blades of the turbine. As the turbine shaft spins, it drives the generator and this produces electricity. The electricity passes along wires to the transformer where the voltage is increased. The electricity then passes into the national grid for distribution to homes, schools, and industry.

Waste gases

The spent steam is cooled, so that it condenses back into water and can be recirculated back to the boiler. The clouds that can be seen escaping from the top of cooling towers are created by steam that escapes and condenses to form water droplets. These clouds are just water. However, the exhaust gases from the boiler leave through a chimney, and they may contain gases, such as CO_2, sulphur dioxide, and nitrous oxide. These are gases that contribute to much of the world's air pollution. Although coal is a very convenient fuel to burn in power stations, it reduces air quality quite significantly. Another waste is the ash left in the boiler. This is used in making concrete and in building roads, although sometimes it is dumped in landfill sites.

Better efficiency

Older designs of coal-fired power stations convert about 35 percent of the energy locked up in coal into electricity. However, newer designs of power stations have improved this. One such method involves the gasification of coal. When coal is brought into contact with steam and oxygen, reactions take place that produce gases such as carbon monoxide and hydrogen. These gases can be used to power gas turbines. Gasification can increase the efficiency to more than 50 percent, as well as produce less solid waste and lower emissions of CO_2, sulphur dioxide, and nitrogen oxides. A process called Integrated Coal Gasification Combined Cycle (IGCC) increases the efficiency further, by using the waste heat from the gases to produce steam to drive a steam turbine, as well as a gas turbine.

CASE STUDY: Coal powers China

China is dependent on coal, because it provides 80 percent of the country's electricity. The rest of its energy comes from either nuclear or hydropower. China's economy is growing at such a rate that it plans to build more than 500 new coal-fired power stations during the next couple of decades. Already, China is the world's second largest producer of CO_2, and this huge expansion in the number of power stations will increase the emissions further. Another problem is that most of China's coal is low-grade brown coal. This type of coal releases sulphur dioxide when it is burned. Unless the power stations are fitted with flue gas desulphurization filters, damage from acid rain also increases.

△ Filling up cars with gasoline or diesel is becoming increasingly expensive. Diesel cars are more fuel-efficient than gasoline cars, since they can run farther on a gallon of fuel.

Burning oil in vehicles

Oil has many uses, but one of the most important of these is in vehicle engines. Engines carry out work; for example, a car engine transforms the energy locked up in a fuel into movement energy. This energy conversion takes place within the car engine, so it is described as an *internal combustion engine*. This contrasts with an external combustion engine, such as a steam engine, in which the fuel is burned outside the engine to produce steam, which then drives the engine.

Four-stroke engines

Most car engines use a four-stroke combustion cycle, invented in 1867. There are four parts to the cycle:
1. The intake valve opens, and the piston moves down to allow a mix of air and fuel into the cylinder.
2. The piston moves back up and squashes the fuel mix. This is the compression stroke.
3. The spark plug produces a spark that ignites the fuel. The fuel explodes and the piston is then forced down.
4. Then the exhaust valve opens and the exhaust gases leave the cylinder.

In the engine, the piston is attached to the crankshaft. The up-and-down movement of the piston turns the crankshaft, which turns the wheels. Modern car engines have four or six cylinders in order to produce enough force to power a car's movement.

Different engines

There are two main types of car engine—gasoline and diesel. The difference between the two engines is the fuel that the engines burn, and the way that the fuel is introduced into the engine and ignited.

In a diesel engine, the compression stroke compresses only air and not fuel. There is no spark in this process. Instead, the compression of the air generates enough heat to ignite the fuel when it is injected into the cylinder. Diesel is heavier and oilier than gasoline. There is more energy locked up in the heavier diesel fuel, so a gallon of diesel produces more energy than a gallon of gas. This means that when diesel burns, it releases about 30 percent more energy than gasoline, so a diesel engine can run farther on a gallon of fuel than a gasoline engine.

Diesel fuel is also cheaper to refine. In the past, diesel engines were powerful, reliable, and economical, but they were also noisy and produced a lot of sooty particles in the exhaust fumes. For this reason, they tended to be used in commercial vehicles, such as vans and trucks, instead of in cars. However, over the last 20 years or so, diesel engines have improved and many cars now have diesel engines.

CASE STUDY:
Liquefied petroleum gas

Liquefied petroleum gas (or LPG) is an alternative fuel to gasoline or diesel. It is a cleaner-burning fuel that produces significantly less air pollution compared to other fuels. LPG contains either butane or propane gases. These gases exist as liquids only at very low temperatures or under pressure. LPG is stored under pressure as a liquid in a tank within the car.

When the pressure is released, the liquid boils and forms a gas. This gas is used to power the car. Car engines powered by LPG fuels are just as powerful as car engines running on diesel or gasoline, but they are very quiet and produce many fewer polluting emissions. LPG is becoming more popular and it is now possible to convert a diesel or gasoline engine to run on LPG.

CHAPTER 4 Environment and resources

Each year, an average-sized coal-fired power station produces about 11.8 million tons (12 million tonnes) of CO_2, 31,300 tons (31,900 tonnes) of nitrogen oxide, 17,322 tons (17,600 tonnes) of sulphur dioxide, 1 million tons (1.1 million tonnes) of ash, 1,000 tons (1,100 tonnes) of dust, as well as 541,000 tons (550,000 tonnes) of gypsum and 22,600 tons (23,000 tonnes) of sludge.

The gases contribute to problems such as acid rain and global warming. Gypsum and ash are solid waste that have to be disposed of safely.

▽ These conifer trees are suffering from acid rain damage. They have dropped their needles and the branches are showing signs of dieback.

Acid rain

Coal is a dirty fuel compared with oil and gas, because it contains impurities such as sulphur. When it burns, the sulphur reacts with oxygen in the air, producing sulphur dioxide—a major contributor to acid rain.

Acid rain has many effects. It erodes stone, especially the limestone and sandstone used in buildings. It also damages trees. Conifer trees are particularly vulnerable, since they tend to grow on thin, acidic soils that are easily damaged by acid rain. The acid rain causes their needles to die and drop off. Acid rain also drains into watercourses, such as streams, rivers, and lakes and makes the water more acidic. This harms aquatic animals, especially fish and amphibians.

Preventing acid rain

Sulphur dioxide emissions from power stations can be reduced by fitting special filters in the chimney to absorb the sulphur dioxide gas. This is called *flue gas desulphurization*. In the United States and European Union, there are strict laws that control the emissions from power stations. Many large power stations, for example, Drax, in England, have fitted flue gas filter systems in order to meet the new regulations. Another way to reduce sulphur dioxide emissions is to switch from burning coal to burning gas, because it produces little sulphur dioxide and even less CO_2.

Carbon dioxide

Power stations also emit CO_2, the greenhouse gas that traps heat in the atmosphere. Scientists believe that warming of the lower atmosphere by just a few degrees could cause enormous climate changes, resulting in more extreme weather events, such as hurricanes, tornadoes, floods, and droughts. When water is heated, it expands. This is called *thermal expansion*. Therefore, a slight increase in the temperature of seawater causes it to expand. Thermal expansion has already raised sea levels by up to 50 inches (20 centimeters) in the last 100 years. This is combined with the effect of melting glaciers and polar ice caps. In time, low-lying areas, especially in Bangladesh, Florida, and islands such as the Maldives, could be flooded.

Damaging the ozone layer

Another emission from burning fossil fuels is nitrogen oxide. About 40 percent of it in the atmosphere comes from burning fossil fuels in power stations and from cars. Nitrogen oxide reacts with water in the atmosphere to produce acid rain, but it also damages the ozone layer. The ozone layer, high up in the atmosphere, acts as a shield from the Sun, filtering out harmful ultraviolet (UV) radiation. Damage to the ozone layer lets more UV radiation through and can cause skin cancers.

CASE STUDY: Drax power station

The largest fossil-fueled power station in Europe is Drax, in Yorkshire, England. Drax has six generators that produce 4,000 MW (4 million KW) of electricity—equivalent to 7 percent of the U.K.'s electricity needs. It is also the U.K.'s cleanest and most efficient power plant. It burns coal and has a desulphurization unit that removes 90 percent of the sulphur dioxide. Drax produces more electricity per ton (tonne) of coal than other power stations because of its advanced control systems. Scientists at Drax are experimenting with burning a mixture of coal and gasolineeum coke to reduce CO_2 emissions. Petroleum coke is a carbon-rich solid formed at the end of the oil-refining process. It releases a lot of energy when it is burned, but it does contain sulphur.

Pollution from transportation

Transportation is powered by oil. Motor vehicles, aircraft, trains, and ships all rely on a liquid fuel derived from oil. When this fuel is burned, it produces a mix of gases including CO_2, nitrogen, water vapor, nitrogen oxides, and sulphur dioxide. The demand for transportation is increasing. In 2006, the world vehicle fleet was about 800 million, of which 37 percent of the vehicles were in the U.S. They are all contributing to the rise in CO_2 levels in the atmosphere.

CASE STUDY: Low sulphur fuels

In recent years, there has been a switch from normal diesel and gasoline fuels to ultra-low sulphur fuel. Ultra-low sulphur diesel contains less than 50 parts per million of sulphur, while normal conventional diesel may contain as much as 500 parts per million of sulphur. This means there is less sulphur dioxide in the exhaust fumes, and the fuel produces fewer sooty particles that blacken buildings, cause acid rain, and are linked to diseases, such as asthma and cancer.

Cleaning up emissions

Catalytic converters can be placed in a vehicle's exhaust system to remove much of the carbon monoxide, hydrocarbons, and nitrogen oxides. Sunlight breaks down the hydrocarbons to form oxidants, which react with nitrogen oxides to cause low-level ozone (O_3), a major component of smog. Nitrogen oxides contribute to both smog and acid rain and are also irritants that can cause human respiratory problems.

Hybrid cars

Catalytic converters do not remove CO_2. The way to reduce emissions of CO_2 is by burning less fuel. In general, a car with a small engine burns less fuel. However, many people are more concerned about a car's performance than its CO_2 emissions.

A different approach is the hybrid car—which is a cross between a traditional car with a four-stroke engine (see page 20) and a car with an electric motor and battery. Hybrid cars are powered by both systems. At low speeds, when less power is needed, the car is powered by the electric motor, but at higher speeds the gasoline engine kicks in. Hybrid engines are more efficient than traditional car engines, so they can travel farther on a gallon of gas. Some hybrid cars are up to 35 percent more efficient than the

△ The Toyota Prius is a hybrid car that is especially popular in California. California has some of the strictest emissions regulations in the world.

equivalent model with a gasoline engine. As a result, emissions are also much lower. The benefit is greatest at low speeds when the electric motor is running, since there are no emissions. This is very important in cities because hybrid cars use their motors in traffic. Hybrids also have advantages over electric cars. The battery in an electric car has to be charged, but the battery in the hybrid car is charged by the gasoline engine, so there is no risk of being stranded with a flat battery.

THE ARGUMENT: We should all be driving hybrid cars

For:
- Hybrid cars have lower polluting emissions.
- They are more fuel-efficient.
- They cause less noise pollution because the motor is almost silent.

Against:
- Hybrid cars cost more than conventional cars.
- A hybrid car's battery needs to be replaced about every ten years.
- They are slower to accelerate than a similar-sized conventional car.

Land use

The extraction of fossil fuels, especially coal, can damage large areas of countryside. Also, more land is lost to new roads and ports that are built to service the new mine or oil field.

Land use and coal mining

Open-pit mining spoils large tracts of countryside. For example, in Germany there are vast open-pit lignite mines and whole villages have been dug up, leaving behind holes in the ground.

A well-planned open-pit operation involves removing the soil and the overburden, and storing them separately. They can then be used to refill the hole when the coal is removed and the area can be restored to its former use or given a new use. Sometimes the land has to be leveled and acidic soils treated with lime and fertilizer so that plants will grow. Former open-pit mines can also be used as landfill sites for garbage.

In some regions, such as the Appalachian mountains, whole mountain peaks have been removed using explosives and the overburden has been pushed into nearby valleys. This flattens the land and gives the miners access to the coal. However, the environmental damage of this operation is widespread; for example, forests are cleared and river courses are changed.

CASE STUDY:
Mining and the Appalachian mountains

There are thick coal seams under the Appalachian range. As a result of rising coal prices and changes in laws, mining companies want to dig more mines in the mountains. However, these companies use a technique known as *mountain peak removal*, and this is having a huge effect on the environment of the mountains. Already, thousands of acres of forest have been cleared, and hundreds of mountain ridges have been blown apart.

Almost 1,240 miles (2,000 km) of watercourses have been buried under piles of rock. Huge slurry ponds threaten local communities. In 2000, a slurry pond in Kentucky spilled 400 million gallons (1,500 million liters) of coal sludge into two mountain streams. All aquatic life in the streams died, and drinking water supplies were affected for hundreds of miles along the Ohio River. Some rivers were so polluted that the water was black for weeks.

The coal is transported to a processing plant where it is washed. The waste from this process is stored in earthen dams containing millions of gallons of slurry. Often these slurry ponds are ignored, with devastating consequences. In 1972, the dam around a slurry pond broke and flooded a valley near Buffalo Creek, West Virginia, killing 125 people.

Deep mines can also scar the landscape. During the mining process, rock from around the coal seam and low-grade coal is brought to the surface and piled nearby to create a spoil heap. These wastes are usually very acidic and contain high levels of metal salts, creating conditions in

▽ Large areas of mangrove swamp have been cleared in Australia to make way for oil storage depots and associated industries.

which plants cannot grow. Any water that drains off this waste may also be polluted. Restoration involves the landscaping of waste heaps as well as treating the acidic soils.

Land for oil and gas

Onshore oil and gas fields can also cause the loss of valuable habitats, although not as much as coal mines. If permission has been given to drill for oil in wildlife areas or other sensitive areas, angled wells may be drilled from outside of the area.

Often the oil is piped to a port for transporting around the world. The building of oil storage tanks and associated buildings can result in the loss of valuable estuarine or mangrove habitats. Also, there is a maintenance zone around the pipelines, which may occupy additional areas of land.

Cost and investment

Fossil fuels have been the main energy source in power stations for more than 100 years. As a result, the technology is well understood and advanced. The cost of the raw material (not taking into account environmental costs) remains relatively cheap.

Fossil fuels versus renewables

Large power stations have a lot of buying power and can use this to make savings. For example, large quantities of fuel are purchased and lower prices can be negotiated. Therefore, the cost of generating electricity from fossil fuels is cheaper than using renewable sources of energy, such as wind, solar, or water. However, the cost advantage of fossil fuels changes when the environmental costs are considered.

> ❝The threat of global climate change and the warnings about energy security will force Europe to drastically change and diversify its sources of supply, relying more and more on renewable energy.❞
>
> European Commission Report, 2004

Using biomass

Some small-scale power stations have been built that use biomass fuels; for example, the waste litter from poultry farms, and wood from fast-growing plants such as willow. Biomass fuels are bulky and their energy content is less than an equivalent mass of coal, so they tend to be difficult and expensive to transport. They have to be used locally to avoid these costs. However, these crops are easy to grow and harvest with little damage to the environment. In fact, willow crops tend to attract many insects and birds. Poultry litter is considerably cheaper than coal, but there is a limited supply, and it is particularly bulky, so the power plant has to be built in areas where there is poultry farming.

Taxing CO_2

Another important consideration is a tax on CO_2 emissions. It is likely that governments will tax industries that release CO_2 to reduce emissions. This could double the cost of generating electricity from fossil fuels, and would bring the costs in line with those of using renewable energy sources.

◁ Power stations, with their cooling towers and coal stores, occupy large areas of land and require good road and railroad communications.

THE ARGUMENT: We should invest in renewables

For:
- Renewables provide a never-ending source of energy.
- They make use of waste materials, such as poultry litter, in the case of biomass power.
- Renewables are ideal for use in local, small-scale power stations or microgeneration in people's homes and buildings. For example, solar panels can be fitted in homes.
- Renewable energy sources are better for the environment, since they give off less CO_2.

Against:
- It costs more to generate electricity from renewable sources, such as water, wind, and Sun, than from fossil fuels.
- A continuous supply of energy from the wind or Sun cannot be guaranteed.
- Biomass crops are bulky. This makes them both difficult and expensive to carry long distances.
- The energy content of biomass is less than that of coal.
- Nuclear power may give off less CO_2, but disposal of the waste and the threat of nuclear weapons make it an unfavorable choice.

Future power stations

As demand for electricity increases, the electricity generating companies need to make the decision to either modify the existing power stations to enable them to generate more electricity, or to build new ones.

Building new plants

Most of the larger power stations have a steam turbine (see page 18), with the fossil fuel releasing heat to generate steam. However, over the last 20 years or so, other designs of power station have been built. The availability of natural gas has meant that many of the newer plants use gas turbines—resulting in fewer emissions.

△ One of the 500 new coal-fired power stations being built in China to meet the country's increasing demands for electricity.

Added to this, the gas turbine is more efficient than a steam turbine that is powered by coal. The cheapest type of power station—and one of the most efficient—is a combined-cycle gas turbine. The hot waste gases from the turbine are used to produce steam that powers a second turbine. This use of waste heat energy increases the efficiency levels to more than 50 percent. In North America and Europe, the cost of this type of power station to supply about a million homes is from $665 to 760 million.

Modifying old power stations

New power stations are very costly to build, so it can be cheaper to use new technology to improve existing power stations, for example, by using a new design of turbine or new mixes of fuel to increase the amount of energy released. Power stations can be cleaned up by the fitting of flue gas desulphurization filters in the chimneys (see page 22). It is also possible to modify a coal-fired power station so that all CO_2 emissions are captured. This process is known as *clean coal technology* and was developed in Australia. The coal is burned in pure oxygen, and the CO_2 is then collected and pumped deep below ground into saline water-bearing rocks or depleted oil fields. However, the long-term effects of this method are uncertain.

Nuclear power stations

Another type of power plant is the nuclear power station, fueled by uranium instead of fossil fuels. The cost to build one of the latest designs of nuclear power station can range from $650 to $1.5 billion—a huge investment. However, the energy yield per ton (tonne) of uranium is much greater than that of fossil fuel. For example, a 1,000-megawatt nuclear power station requires about 163 pounds (74 kilos) of fuel per day; but an equivalent-sized, coal-fired plant needs about 9,310 tons (9,460 tonnes) of coal per day. There are no CO_2 emissions from nuclear energy, but radioactive waste must be stored safely for hundreds of years. For this reason, many countries that have the technology for nuclear power prefer to invest in renewable energy sources.

THE ARGUMENT: We should build new fossil fuel-powered power stations

For:
- Power stations provide a reliable source of electricity for domestic and industrial use.
- They provide jobs.

Against:
- Clearance of land for new power stations and coal stores, approach roads and railroads causes loss of wildlife habitats.
- The power stations create air pollution from burning coal.
- They create solid waste that has to be disposed of.
- As fossil fuels are depleted and their cost rises, an investment in new power stations may not be profitable.

Transporting fossil fuels

Once the fossil fuels have been extracted from below the ground, they need to be transported to power stations and oil refineries.

Moving coal

Coal is a bulky fuel so it is usually moved by train across land. In many places, power stations are built close to the coal mines so that the distances are short. However, some countries lack any coal reserves, so their coal supplies are imported by train or sea.

Moving oil

Oil is usually piped to a collecting area and from there it is either moved into a network of pipelines, or transferred to large oil tankers and moved by sea. There are about 3,500 oil tankers currently in operation, the largest of which are equipped to carry up to around 492,000 tons (500,000 tonnes) of crude oil.

Oil is harmful if it is spilled into the environment. Since it is a liquid, it can seep into watercourses. The loading and unloading of oil at ports is a risky operation. Most ports use barriers and other methods to minimize any spill—should it occur. However, the greatest risk is the oil tanker having an accident at sea, causing large amounts of oil to pour into the sea. Some oil tankers are fitted with a double hull.

This means that if the outer hull is damaged, the inner hull protects the ship and also prevents an oil spillage. However, tankers have a very long lifespan, and there are still some single-hulled oil tankers in use that were built in the 1960s and 1970s. Fortunately nowadays, their access to some coastal waters, such as Prince William Sound in Alaska, is restricted.

Moving gas

Gas is usually piped from the gas fields to the end users via various storage facilities. Some of the gas pipelines are long; for example, some pipelines run all the way from Central Asia to Europe. As natural gas becomes an increasingly important fuel in the world today, more pipelines are planned. In 2002, the government of Afghanistan signed an agreement to build a 530-mile (850-kilometer) pipeline that would carry gas from Turkmenistan to Pakistan and India.

" Modern society will find no solution to the ecological problem unless it takes a serious look at its lifestyle. "

Pope John Paul II, 1990

△ Cleaning up an oil spill. Unsightly as it is, some scientists argue that it is best to leave the oil and let natural processes break it down.

CASE STUDY:
Exxon oil spill

One of the worst oil spills occurred in 1989 when the oil tanker *Exxon Valdez* hit a reef outside Valdez in Alaska, spilling its oil into Prince William Sound. Three days later, a huge storm pushed the oil farther afield, affecting 1,200 miles (1,900 km) of Alaskan coastline. Exxon claims that 1.5 million cubic feet (40,900 cu m) of oil were spilled in the accident, but environmental groups believe the oil spill to be up to 3.9 million cubic feet (110,000 cu m). A huge clean-up operation took place, but not in time to save the lives of many thousands of animals, including about 250,000 sea birds, 2,800 sea otters, 300 harbor seals, 250 bald eagles, up to 22 orcas, and billions of salmon and herring eggs. Although signs of the oil spill were gone within a year, the effects can still be felt. There has been a significant decline in the numbers of marine animals, such as sea otters, ducks, and pink salmon.

Exploring for oil and gas

Most of the world's large oil and gas fields have been discovered. Now oil companies are looking for new reserves in increasingly remote areas, such as the Arctic, the Sahara, and beneath some oceans. The oil company BP is currently exploring in Alaska, while the Premier Oil company is exploring for oil beneath the Western Sahara.

Although costs are high, oil companies are investing billions of dollars into exploration and production. It is estimated that up to $800 billion will be needed by 2013 to develop existing reserves as well as exploration costs, estimated at about $240 billion a year. For example, offshore wells are very expensive to drill. In 2005, it cost about $37,000 per day to rent an offshore rig to drill a new oil well. It usually takes about 45 days to drill a new oil or gas well, so it can cost up

△ Carrying out a geophysical survey in the Sahara to locate new oil fields.

to about $1.7 million. Following this, a production platform is brought in, which costs around $9 million or so.

> ❝ Developing a small section of Arctic National Wildlife Refuge would not only create thousands of new jobs, but it would eventually reduce our dependence on foreign oil by up to a million barrels of oil a day. ❞
>
> President George W. Bush, 2005

New oil fields

Once a new oil reserve is found, it can take several years or more to bring the first oil to market. The oil field has to be mapped, test wells drilled and then drilling rigs brought in. Then pipelines and storage facilities need to be constructed. Finally, the oil has to be shipped to the oil refinery.

Some of the new oil fields may never be developed because of the high costs involved in extraction. Even if the price of oil rose, there would be an accompanying increase in the costs of plastics, steel and transportation. Although the oil underground may have increased in value, the cost of extracting it has also increased.

There are many other reasons why oil reserves are not exploited. For example, the state of Florida has oil reserves, but they have not been developed because people are concerned that oil production is not compatible with tourism. Sometimes there are environmental reasons why the oil cannot be extracted.

One of the most controversial oil fields lies under the Arctic National Wildlife Refuge, a huge wilderness area in northern Alaska. It is one of the most important wildlife areas in the Arctic. However, the Refuge lies to the east of Prudhoe Bay, the largest oil field in the United States, and research has shown that substantial oil reserves lie beneath the Refuge. This pristine environment is under threat from the oil industry. The final decision as to whether to extract the oil has to be made by the U.S. Senate.

THE ARGUMENT: The U.S. should exploit oil from the Arctic National Wildlife Refuge

For:
- Directional drilling would allow oil to be extracted by disturbing only 1 percent of the refuge.
- Less than 0.01 percent of the land area would be affected.
- Drilling in the Arctic would reduce U.S. dependency on foreign oil.
- Hundreds of thousands of jobs could be created directly or indirectly.

Against:
- The U.S. should instead spend its money on renewable power and combating CO_2 emissions.
- Drilling in the Arctic could make it easier for oil companies to exploit other wilderness areas.
- The drilling would interfere with the herds of caribou on which people depend for food.
- Any oil spill would severely damage the environment, and its effects would be long lasting.

Alternative fossil fuels

Coal, gas, and oil are the main forms of fossil fuels, but they are not the only ones. There are other fossil fuels, such as peat, tar sands, and oil shales. With the right technology, these could be put to greater use in the future.

Peat

Peat is a soft, brown material made from compressed and partially decomposed vegetation. It forms in a waterlogged habitat, such as a swamp or bog. It has about 50 percent carbon content, and can be dried and burned as a fuel. However, it does not burn as efficiently as coal.

One of the main problems with peat is the loss of valuable wetland habitats known as *bogs*. In order to extract the peat, the bog is lost. Hand-cutting of peat is a slow and labor-intensive process, but it can allow the bog to recover partially. In contrast, large-scale mechanized extraction processes drain and strip all vegetation from vast expanses of bog, often completely destroying the habitat.

Oil shales

Oil shales are rocks that are rich in an organic material called kerogen. They can be burned as a poor quality coal, but they are usually processed to produce oil. They are crushed and heated from 830°F to 930°F (445°C to 500°C) without oxygen. This is called *pyrolysis*. The kerogen within the rock

▽ Peat is dug by hand and left to dry, then it is sold locally as an alternative fuel to coal.

CASE STUDY:
Oil from coal

During World War II (1939–1945), Germany developed a technique for extracting oil from coal, because there was a shortage of crude oil. Today, a leader in this field is South Africa, a country with rich coal reserves but little oil. The coal-to-oil process has been modernized by the oil company Sasol. Although this process could supply large quantities of oil, it is not without its problems. Coal-to-oil fuels become cost-effective only when oil prices exceed $37 a barrel (160 liters).

However, the coal-to-oil process is also environmentally damaging, because considerable CO_2 is released during the production of the fuel, as well as when it is burned. In fact, twice as much CO_2 is produced than is released from the same amount of gasoline derived from crude oil. Coal-to-oil may become more common in the future, and the benefits of this new source of oil must be weighed against the increased emissions of CO_2 and how this affects global warming.

breaks down and forms oil. However, the mining and processing of oil shale is an energy-hungry process—up to 40 percent of the potential energy in the shale is used in oil production.

The world's oil shale reserves could yield up to 105 trillion gallons (400 trillion liters) of oil. Of this, about 40 trillion gallons (150 trillion liters) are located in the U.S. Oil shales are mined in Estonia, Russia, Brazil, and China, but until recently, mining was in decline because of the economic and environmental problems associated with open-pit mining. Also, the extraction of oil produces large amounts of ash and waste rock, which

have to be carefully disposed of. However, as oil prices rise, the large-scale mining of oil shale may become more economically viable.

Tar sands
These are sands that contain bitumen, a heavy oil. The largest deposits in Canada may contain 66 trillion million gallons (250 trillion million liters). However, extracting the oil is costly and damages the environment. Surface sands are dug from open-pit mines and crushed to extract the oil. Deeper deposits are extracted by pumping steam into the ground to create a slurry. At the extraction plant, the bitumen is skimmed off the top.

CHAPTER 6

A fossil fuel-powered future?

How long will the fossil fuels last? Everyone agrees that there is a limited supply of coal, gas, and oil, although not everyone agrees about how much of them are left. However, it is clear that we have to find ways of making fossil fuels last longer.

How much gas?

Global reserves of natural gas at the end of 2005 stood at 6,300 million million cubic feet (180 million million cubic meters), of which 75 percent lie in Europe, Russia, Central Asia, and the Middle East. The reserves have increased by 25 percent since 1995, but consumption has increased by a similar percentage. Natural gas production is likely to peak around 2030. After this, some say, it will reach maximum output and then start to decline. In the future, other sources of natural gas, such as landfill and organic wastes, could be exploited.

How much oil?

Oil reserves have increased by more than 50 percent in the last 22 years and are expected to continue to increase as more oil is discovered. In 2006, global oil reserves were estimated at about 1,300 billion barrels, of which about two-thirds lay in the Middle East and Canada.

There are many estimates of when peak oil production will be reached. A French government report suggested that peak oil production will be in about 2013. This is based on a 3 percent increase in demand each year and discoveries of new oil fields with reserves of 10 billion barrels per year. Oil consumption per day would reach a maximum of around 97 million barrels per day. Other estimates suggest that the peak will be later, somewhere between 2020 and 2030.

How much coal?

Coal resources are estimated to be ten times larger than oil reserves, and may last several hundred years. It is also thought that the large quantity of coal waste that is stored near the coal mines could become exploitable with the use of new technology.

Saving energy

Fossil fuels could last much longer if they were used carefully. The efficiency of power stations could be increased so more electricity was produced from every ton (tonne) of fossil fuel. Consumers could reduce their consumption of electricity by using low-energy appliances. Car engines could be more efficient, so that they traveled farther on a gallon of fuel,

△ Fuel cell-powered buses are appearing
in many cities such as Stockholm, Sweden.

and efficient, cheap, reliable public
transportation systems would encourage
more people to leave their cars at home.

An example of energy savings is
happening in Southampton, England,
where waste heat from power
generation is used to heat buildings.
The city has integrated a geothermal
source of hot water, with a combined
heat and power generator producing
5.7 megawatts of electricity, as well
as hot water. Many thousands of
buildings in the city are connected
to the district's heating scheme.

> **“** What is very important is
> conservation, especially in
> transportation. Raising taxes on
> fuel, introducing toll roads and
> bridges into major cities, for
> example, but also stopping the
> spread of suburbs ever further
> from city centers. Controlling
> suburban blight is one way to
> slow oil consumption until we are
> a society no longer dependent
> on oil. **”**
>
> Deborah White, Senior Energy analyst, Société
> Générale, Paris

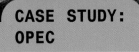

The Organization of the Petroleum Exporting Countries (OPEC) is an intergovernmental organization set up in 1960. It has many aims, which include:

- the coordination of oil policies of the member states
- stabilization of oil prices in order to secure a steady supply of oil to customers
- provision of a regular and fair income to the oil producers and others that invest in the industry.

OPEC's members control about two-thirds of the world's oil reserves and supply 40 percent of the world's oil production. Decisions made by OPEC regarding the volume of oil being produced have had a considerable influence on world oil prices. For example, in 1973, OPEC started an oil crisis when its members refused to ship oil to those

△ Representatives of the 11 members of OPEC meet regularly to consider changes in the world's oil markets and ensure that there is balance between output and demand.

countries that had supported Israel in their war against Egypt and Syria. Oil prices rose and there were gasoline shortages in many MEDCs. This lead to considerable investment in renewable energies, for example, in the United States there was a great expansion of wind farms and solar power plants. However, interest in renewables waned when the oil prices dropped back again. The surge in prices since 2004 has led to new interest in renewables. During the last ten years or so, the influence of OPEC has decreased as oil production from non-OPEC members has increased; for example, oil can now be supplied from the Gulf of Mexico and the North Sea.

THE ARGUMENT: Governments should control oil prices

For:

- By pricing their own oil, governments are given the ability to price oil so that it suits their country's current economic situation.
- Governments can increase oil prices to reduce consumption.
- If governments price oil, they can prevent oil companies from making large profits.

Against:

- Oil can be used as a political tool. For example, by reducing the oil price, governments would be more popular with voters, especially as energy prices rise.
- If governments made oil cheaper, this would reduce the investment in alternative energy sources.

Oil and politics

Governments are starting to plan for a future where coal, gas, and oil may not be so readily available and will cost much more. Forward planning is essential, because it can take decades to build new power stations and establish alternative energy sources.

The rising cost of oil

During 2005 and 2006, the price of oil rose to record highs. There were many reasons for this. Oil prices usually rise during times of instability in oil-producing countries, because there is a fear of supplies being cut. However, the main force that drives the cost of oil is demand. It has been estimated that world oil demand is set to increase by 47 percent between 2003 and 2030. Much of the increase is due to the fast-expanding economies of India, China, and other Asian countries. However, the supply of oil cannot keep up since most oil producers are working at full capacity. Only Saudi Arabia has spare capacity to increase its production. Also, there is an increased demand for the high quality, light crude oils that are used to produce the clean-burning fuels used in cars, whereas most of the spare capacity is for heavier oils.

> **What we human beings are all living now... is an extraordinary but exceptionally dangerous adventure. We have a very small number of years left to fail or to succeed in providing a sustainable future to our species.**
>
> Jacques Cousteau, French underwater explorer and author

Securing supplies

Many regions have no natural supplies of fossil fuels of their own, so they are reliant on supplies from other countries. For example, the EU is a net importer of energy, and by 2020, about two-thirds of the EU's total energy requirements will be imported. This means that up to 75 percent of its natural gas supplies could be imported from countries such as Russia.

Countries with large reserves of oil and gas have a lot of political weight. By increasing or decreasing the supply of fuel, they can influence the world economy. For example, Russia has huge reserves of coal, gas, and oil. Already people have accused Russia of using energy as a political tool to further its influence, especially with countries of the former Soviet Union.

In January 2006, a dispute over gas prices led to the Russian gas company cutting off gas supplies to the Ukraine. The Russians wanted to increase the cost of gas from $45 to $200 per 35,300 cubic feet (1,000 cubic meters), claiming this was the market price. The Ukraine refused and demanded a phased price rise. Many believed that this was political because of the Ukraine government's close links with the EU. In contrast, Belarus, another former Soviet state, continued to be supplied with gas at $45 per 35,300 cubic feet (1,000 cubic meters). Around the world, many countries are looking to secure alternative supplies to reduce the risk of interruptions to their oil and gas supplies.

New pipelines

In 2006, the new Baku-Tbilisi-Ceyhan crude oil pipeline was opened. It runs 1,000 miles (1,700 kilometers), from Baku, Azerbaijan, through Georgia and to the port of Ceyhan in Turkey.

A new gas pipeline is also planned. The 2,000-mile (3,300-kilometer) Nabucco gas pipeline will run from Turkey to Austria. Construction is expected to start in 2008 and be completed by 2011, at a cost of over $4 billion. It will pipe up to 1,000 billion cubic feet (30 billion cubic meters) of gas from countries such as Iran, Azerbaijan, and Turkmenistan. These oil and gas pipelines will reduce the EU's dependence on oil and gas from the Middle East and Russia, and will provide a boost to the economies of Azerbaijan, Georgia, and Turkey.

Alternative oil supplies

Some countries are turning to alternative oil sources. Oil production using Canadian tar sands in Alberta has increased by 60 percent since 2002. Oil from tar sands is difficult to extract, but just extracting the most accessible 10 percent would yield 7.4 trillion gallons (28 trillion liters) of oil.

△ During the Gulf War (1990–1991), the retreating Iraqi army set alight many oil wells in Kuwait. This created clouds of thick, black smoke.

CASE STUDY:
Terrorist targets

Any interruption in supply of gas or oil to a country could have severe consequences on that country's economy and political leadership. Therefore, it is not surprising that oil and gas pipelines are quickly becoming terrorist targets. Pipelines are vulnerable to attack because they are difficult to protect. They run for thousands of miles and cannot be protected along their entire length. Some pipelines are built underground. This offers some protection, but there are vulnerable points above ground. Already pipelines in the Caucasus, Nigeria, Colombia, Sudan and Yemen are regularly blown up, while oil facilities have been attacked in countries such as India, Pakistan, Indonesia, Argentina, and Ecuador. In the country of Georgia, the new oil pipeline is protected by patrolling antiterrorist units, but they will be unable to protect the pipeline from missile attacks from Iran. An attack could lead to huge oil spills that could harm wildlife and contaminate water supplies, as well as interrupt supplies to the EU.

Looking to the future

Fossil fuels are part of our modern lives. However, the supply is limited and the world will have to become less dependent on fossil fuels. New discoveries, the mining of oil shales and tar sands, and improved fuel efficiency will all help to extend the life of fossil fuels. Although experts disagree as to when these fossil fuels will run out, they are all agreed that they will run out, eventually. Before this happens, other sources of energy, such as the Sun, wind, water, and nuclear power, will have to be developed to take the place of fossil fuels. It is going to take decades to develop these energy sources, so it is imperative that this process starts soon.

Combating global warming

It is not just the limited supply that will force people to shift away from fossil fuels. Fossil fuels are a major contributor to global warming. If we are to fight climate change by reducing CO_2 emissions, we will have to start applying clean coal technology, use renewable energy sources, and reduce the amount of oil burned by the millions of vehicles on our roads.

Making difficult decisions

Combating climate change will mean making difficult and, possibly, unpopular decisions. For example, to reduce the polluting emissions from

" Because renewables do not use fossil fuels [most are entirely fuel-free], they are largely immune to the threat of future oil or gas shortages and fossil fuel price hikes. For the same reason, because most renewable technologies require no combustion, they are much kinder to the environment than coal, oil, and natural gas. Smog and acid rain could be eliminated with renewables. The collective lungs of America could breathe a sigh of relief. "

John J. Berger, environmental science and policy consultant

vehicles, governments may decide to place heavy taxes on larger makes of car. California has already legislated that any cars driven in the state must meet strict emission controls, and it is very likely other states and governments may follow this lead.

In order to generate enough electricity to meet the increasing demand, large areas of land or sea may have to be covered by wind turbines or huge tidal barrages, built across their estuaries.

Going nuclear?

It is also possible that governments will build new nuclear power stations. Although nuclear power does not generate any greenhouse gases, it does create radioactive waste that will need to be stored for hundreds of years.

Scientists are currently developing new designs of nuclear power stations that will recycle their own radioactive waste. However, the designs are still experimental, and it may be some time before a commercial nuclear power station of this design could be built and come into operation.

Governments can legislate to bring about change, but the ordinary person also has an important role to play. If everybody reduced their electricity consumption by just a few percent, or used public transportation instead of their car just once a week, the overall effect would be significant. Action needs to be taken now. Perhaps it will take a major climate crisis, such as global warming, to make people take action and change their lifestyles?

▽ Although hydroelectric power is a sustainable source of energy, the construction of dams has a considerable impact on the environment.

Acid rain Rain that has a pH of less than 5.6, due to the formation of sulphuric, nitric, and other acids.

Asphalt A sticky, dark substance produced from crude oil residues.

Barrel The unit for measuring oil. One barrel is 50 gallons (159 liters).

Combustion The process of burning in air; it takes place in car engines.

Compression Making something smaller by applying pressure.

Distillation Heating a liquid to its boiling point, then condensing it and collecting the vapor.

Emission The release of a substance into the environment.

Exhaust The hot gases released from an engine as waste products.

Fossil fuel The fuels peat, coal, crude oil, and natural gas.

Gasification The process by which coal is turned into a gas.

Generator A machine that produces electricity.

Global warming The gradual warming of the Earth's surface.

Indigenous Peoples or plants that are native to a particular area.

Microscopic Too small to be seen. Only visible using a microscope.

Motor A machine that converts electrical energy into mechanical energy.

Nonrenewable Resources that cannot be replaced or replenished, for example, the fossil fuels coal, oil, and gas.

Overburden The layer of rock and soil that covers a mineral deposit, such as a coal seam.

Peak production A point in time when oil-producing countries reach their maximum output; after this time production starts to decline.

Refinery An industrial plant that processes crude oil or metal ores.

Renewable Resources that are able to be replaced or replenished.

Sediment Particles of sand, silt, and clay that are carried by water and settle in a pond, river, lake, or sea.

Slurry Water or liquid that is full of suspended particles.

Smog A visible haze that forms over a city due to vehicle air pollution.

Sustainable Resources or energy sources that can be maintained at a steady rate for future generations.

Trillion A million million.

Turbine A bladed wheel, turned by moving water or steam, which turns a generator to produce electricity.

Books to read

Science at the Edge: Alternative Energy Supplies Sally Morgan; Heinemann Library, London, 2002: a thorough look at renewable energy supplies.

Science at the Edge: Global Warming Sally Morgan; Heinemann Library, London, 2002; discusses how global warming is affecting our planet.

Sustainable Futures: Energy John Stringer; Evans Brothers, London, 2005; looks at how energy will need to be used if we are to develop a sustainable future.

True Books series Christine Petersen; Children's Press, Danbury, 2004; a detailed look at the different sources of energy and where they come from.

Web Sites

Due to the changing nature of Internet links, The Rosen Publishing Group, Inc., has developed an online list of Web sites related to the subject of this book. This site is updated regularly. Please use this link to access the list:
www.rosenlinks.com/ted/oil/

Index

Note: Page numbers in *italic* refer to illustrations.